I0040215

#BUSINESS SAVVY PM **tweet** Book01

Project Management Mindset, Skills, and Tools for
Ensuring Powerful Business Results

By Cinda Voegtli

THiNKaha®

E-mail: info@thinkaha.com
20660 Stevens Creek Blvd., Suite 210,
Cupertino, CA 95014

Copyright © 2011 by Cinda Voegtli

All rights reserved. No part of this book shall be reproduced, stored in a retrieval system, or transmitted by any means electronic, mechanical, photocopying, recording, or otherwise without written permission from the publisher.

Published by *THiNKaha*®, a Happy About® imprint
20660 Stevens Creek Blvd., Suite 210, Cupertino, CA 95014
http://thinkaha.com

First Printing: August 2011
Paperback ISBN: 978-1-61699-062-6 (1-61699-062-7)
eBook ISBN: 978-1-61699-063-3 (1-61699-063-5)
Place of Publication: Silicon Valley, California, USA
Paperback Library of Congress Number: 2011929747

Trademarks

All terms mentioned in this book that are known to be trademarks or service marks have been appropriately capitalized. Neither Happy About®, nor any of its imprints, can attest to the accuracy of this information. Use of a term in this book should not be regarded as affecting the validity of any trademark or service mark.

Warning and Disclaimer

Every effort has been made to make this book as complete and as accurate as possible. The information provided is on an "as is" basis. The author(s), publisher, and their agents assume no responsibility for errors or omissions. Nor do they assume liability or responsibility to any person or entity with respect to any loss or damages arising from the use of information contained herein.

Advance Praise

"This book provides fast insights that can help lift project managers out of a purely operations mode of thinking—and into strategic business-driven thinking that will truly set them apart!"

Laura Erkeneff, CEO, Training For Techies, Inc.

"Keeping within scope, time, and cost constraints is no longer sufficient to measure project success. This book provides great guidance for project managers who are looking to successfully lead projects that will have a lasting, positive impact on their organization."

Kent J. McDonald, Business Advisor, Knowledge Bridge Partners and co-author of *Stand Back and Deliver, Accelerating Business Agility*

"The practical project management tips that Cinda Voegtli is known for, in small bite-sized nuggets—delicious! I'll use this book as a quick regular reminder to keep my networking projects on track."

Jeff Richardson, Empowered Alliances

Dedication

I dedicate this book to my family, who has watched and supported the varied demands of my evolving career all these years; to my colleagues at ProjectConnections.com who have shared the joys and challenges of managing projects, running a business, and learning to always keep customer needs and business goals at the forefront of every project; and to all of us who are so busy every day that nuggets of wisdom can be a useful way to remind ourselves of what's important!

Acknowledgments

My thanks go out first to Ori Kopelman, a long-time colleague who first powerfully introduced me to a single-minded business-driven project focus. The Project Vision document I use to this day—which gets an entire team truly understanding and aligned to the business reasons for a project—changed my life as a project manager.

To Laura Erkeneff, thanks for your own perspective on bringing a business perspective to technical team members' work on projects, as well as helping me pull together these key concepts, tips, and tools into a great set of ideas, refreshers, and reminders for all of us.

There are numerous other people who have influenced my thinking about the project manager role over the years. Thanks to Rita Glynn Smith for her views on how project managers are just another 'seat at the table' to many functional groups, and how project managers need to act and contribute and lead to be respected by all.

Thanks to Bob Drazovich for his work in our business (and reminding me occasionally to "eat my own dogfood" and get the business back at the forefront of our own projects).

Thanks to all the executives I've worked for and with, for sharing their perspectives and working with me and other project managers to effectively meld our world with theirs—enabling the success of our projects, and the overall benefit of the businesses those projects are meant to serve.

Why Did I Write This Book?

I learned project management the hard way—and not just the tools and techniques of managing tasks and timelines. After getting mixed signals about the project manager's role from a series of initial experiences, I learned how to fulfill the most important responsibility of all—making sure that every project actually delivers meaningful results for the business. Given time and resource constraints so many projects face, it's so easy for goals to get muddied and projects to get off track, with lots of hard work unfortunately followed by marginal benefits to the business! My goal in creating this book is to help save others some of that time and pain! I hope it will communicate clearly a business savvy, business-driven view of the project manager's role, by crisply conveying the mindset, actions, and tools a project manager can use to be a highly-valued, results-producing leader of every effort they undertake.

CindaVoegtli and GreatPMs
cinda@projectconnections.com
blog.projectconnections.com/cinda-voegtli.html
http://www.ProjectConnections.com

Contents

Section I

The PM's Role and Relationships—Being "Business Savvy"

What does it mean to be a "business savvy" project leader? Understand how to be an advocate for the project, from the perspective of its end customers and its business value to your company. Focus on clarity and communications with everyone—the project sponsors, your team, and your customers—with the business as the context. Establish yourself as the leader of a team striving for meaningful business results.

1

The project manager's job:
Establishing, Quantifying, Propagating,
and Protecting the Business Goals of
the project.

2

Establishing the goals: Take the
initiative from the beginning to help
set up the project for business success:
You are not just an order taker!

3

Insist on knowing the project selection criteria & process that led to this project. Why does it exist?

4

The team's job is to provide the right deliverable to customers when needed to fulfill customer needs & meet company financial goals...

5

...at the project cost we are ready to spend and a price they are willing to pay, while leaving the company with a viable profit margin.

6

Systematically consider other impacts of the project on the company (e.g. impacts on system architecture & future development) as well.

7

The business savvy PM
continually insists on
extreme clarity in both the
business goals and the
project justification.

8

Once you have extreme clarity, continually ensure alignment of all the work and all team roles to the business goals.

9

Strive to understand what both the customers and the company need— then lead the team to define and deliver the needed results.

10

The PM leads the way in finding the best balance between time, costs, and features for both the company & the customer.

11

Team members must understand that tradeoffs may need to be made to protect the most important business goals.

12

A successful PM knows that consensus is not always possible: The goal is a project definition that "everyone can live with & support."

13

The PM's role is to drive for balance among functionality and design while leading the team to deliver on time and within budget.

14

Business results are influenced by the executives who set corporate strategies and business goals.

15

Business results are influenced by the managers who decide resources, oversee technical decisions, & spend development funds.

16

Business results are influenced by the team members who create products, make tradeoff decisions, & meet schedules to get to customers.

17

Business results are influenced by all the functions involved in selling and supporting the organization's products or services.

18

Business savvy PMs work proactively across groups to drive the right priorities, the right decisions, and the right overall results.

19

Business savvy PMs work proactively across groups to minimize disconnects and costs and maximize delivery speed & profitability.

20

True project leaders take action—unbidden and with a mature attitude—operating with an executive-level mindset.

21

Powerful PMs drive an effort with a combination of ownership, initiative, and leadership to keep things moving on the right path.

22

They use business language to lead through the craziness, provide execs with information, and push back on unreasonable demands.

23

They do what's needed in the face of tough project trade-off decisions, speaking up with courage and with business impact.

24

The management aspects of the PM role are still important—planning, budgeting, organizing, staffing, controlling, and problem solving.

25

Additional leadership aspects are critical: Guiding direction, championing the goals, aligning people, motivating, and influencing.

26

To effectively manage the team, understand individual styles, and how to best communicate, listen, motivate, reward, & resolve conflict.

27

To effectively manage the work, pay attention to time management, delegation, & organization for yourself and the team.

28

To effectively lead the team, focus on helping everyone do the right things and stay oriented toward the critical business goals.

29

Business savvy PMs understand the business & market drivers, company goals & priorities, and customer goals & priorities.

30

Business savvy PMs are able to credibly drive project trade-offs and make valued recommendations to executives, sponsors & others.

31

Business savvy PMs do not just manage a list of tasks. They manage all the work in the context of the business.

32

A business savvy PM is not a bureaucratic status collector or paperwork coordinator!

33

The business savvy mindset means continually clarifying & aligning the project definition to recognizable business goals.

34

The business savvy mindset means judging the project's progress and success based on continued alignment to business goals.

35

When you show up with a business savvy attitude, you build and enhance your career opportunities: Executives notice those PMs.

Section II

Business Savvy during Project Start-up

It is the business savvy PM's job to understand the business value of a project—and why your company is undertaking it. Only then can the PM emphasize to the team what they are trying to achieve together and why, once a full project is underway. Understand (even challenge!) the reasons for the project, and be able to communicate the company's strategy and how this project fits in. Participate in early consideration of the business ROI and other customer-related or strategy-focused goals.

36

A great PM understands the purpose of the project from the very beginning.

37

Projects get started due to executive mandate, customer demand, or a departmental priority. But does it really make sense?

38

Ask yourself right away, "Should this project even exist? What do I think? How can I tell?"

39

Ask yourself, "What do I need to know up front?" & insist on asking tough questions to find solid answers on this project's purpose.

40

Does this project fit logically into the company's portfolio and make strategic sense?

41

Both subjective assessments &
quantitative ratings can be used to
compare projects and make judgments
of project value.

42

Typical project justifications include
a strategic need, a competitive threat,
regulatory mandates, desired returns,
and cost savings.

43

Typical quantitative measures of a project's value include profitability, return-on-investment (ROI), and net present value.

44

ROI is determined by the profit (or cost savings) to be gained, divided by the cost-of-development.

45

ROI should be estimated roughly before a project begins to "sanity check" the economic viability of the proposed project.

46

Take the initiative: An early ROI can be assessed using early sales estimates from Marketing.

47

Take the initiative: An early ROI can be assessed using early cost and profit estimates from concept work by Development.

48

It's a business savvy PM's job to ensure that ROI continues to be tested as the project moves forward.

49

The concept phase should include the project proposal and a fast go/no-go sanity check. Has one been done for your project?

50

A new project proposal should state the idea, what is needed and why, key business goals, and rough sizing of the envisioned project.

51

Then do a first business case and feasibility assessment—followed by a rough scope recommendation, if this project should go forward.

52

The business case states the next level of justification detail & includes an economic (quantitative) rationale for the project.

53

The kickoff phase will then look further at requirements & solutions, the business case, tradeoffs, & definition of a viable project.

54

80% of the project and product costs are determined in the first 20% of the project's schedule—don't skimp on the front end!

55

A business savvy PM drives the right evaluation to be done in a timely manner, questions cost & benefits, and ensures approval to go.

Section III

Business Savvy during Project Kickoff and Planning

A business savvy project manager gets everyone on the same page with a common and compelling set of expectations and goals for project success. Create a clear vision of what this project must achieve and rally everyone around it—team members, stakeholders, and the executives who are sponsoring your project. Continue to communicate the goals as the project's reason for being and driver of all detailed plans.

56

A strong kickoff means a team together defines a project to be on target with business goals—avoids profit-killing later surprises.

57

A business-oriented kickoff includes team formation, a project vision, alternative reviews, economic justification, tradeoffs, & plans.

58

Avoid project definition problems that impact critical financials. Involve all the right people to avoid "lopsided" project goals.

59

Core teams are cross-functional because business savvy PMs realize that all functional work affects a project's ultimate profit.

60

Get the right team & executives involved & enlisted up front, focusing everyone right away on what the customer needs most.

61

Do ongoing cost-benefit analysis as the project is fully defined; keep quantifying, asking questions, & making recommendations.

62

Identify the core team, the sponsor, stakeholders, & influencers; then involve and engage them in the kickoff process.

63

Hold a kick-off meeting to discuss business goals, draft a project vision, and identify possible alternatives and trade-offs.

64

A kickoff meeting provides a fast, focused start to the project with strong cross-functional and sponsor involvement.

65

Engage your sponsor at the meeting to express and champion the business goals, provide direction, & resolve conflicts.

66

To motivate a Sponsor, understand the WIIFM (What's In It For Me?) for the sponsor and engage them in "business benefit" terms.

67

A two-page project vision document focuses first on customer needs and benefits to be provided, & how they'll measure value received.

68

The project vision document then covers key features, other critical factors, and financials (dates, sales targets, cost targets, etc.)

69

Use the project vision to help clarify that customers are not all the same and may have different requirements and priorities.

70

The vision document serves as an alignment & communication tool, as well as a contract for the project after tradeoffs are made.

71

Use a Project Flexibility Matrix to clarify what can "give": is scope, schedule, or cost most flexible or inflexible?

72

Nuances of what is important may not be fully understood until trade-offs are forced by constraints—that conflict is good!

73

The PM must be able to speak up and willing to "jump in front of the train" if the project definition is on the wrong track.

74

A benefit of gaining
team-wide alignment during
kickoff: no more dictation
or premature quoting of
release dates.

75

Another benefit of the kickoff process: Everyone understands the reasons behind any tough tradeoffs made.

76

Kickoff phase investigation & project tradeoff negotiations are a powerful source of team contribution, understanding, and buy-in.

77

The kickoff process helps avoid "project killers" such as taking one person's word on key goals and later finding out they were wrong.

78

The kickoff process helps cut through the sense that "All goals are important so we must do it all at once."

79

Business goals are not just for Marketing and Executives. The team must fully understand as well.

80

An effective kick-off process yields total company commitment to a concrete, well-defined project vision that meets key business goals.

Section IV

Business Savvy during Project Execution

Great project managers are business minded and profit aware. They maintain a "life-cycle" perspective on the project. They keep in touch with all aspects of the project, including cross-functional groups and business sponsors. The key to keeping the project running smoothly is early identification of issues and problems that could threaten the project meeting its business objectives. Establish the communications channels that provide you with insight and understanding of the project status—related to not just schedule and costs, but also to achievement of business goals. Keep those objectives in mind in making changes or addressing project issues.

81

The business savvy PM
ensures cross-functional involvement
and communication throughout
the project.

82

The goal is to get early
cross-functional buy-in to project
goals and tradeoffs, then stay
engaged throughout the project.

83

"Doing cross-functional right" ultimately means avoiding profit-killing misunderstandings on costs, specs, time, responsibilities.

84

"Doing cross-functional right" means getting the right team members involved at the right time to be involved in reviews.

85

Successful core teams have cross-functional cooperation as early as possible in the project.

86

Late changes in the project hit the bottom-line profit the hardest. Move changes to the front end where they are cheapest.

87

Consider the entire lifecycle of the project & all support groups. Everyone's work & decisions can impact "profitability" of the result.

88

Get the right involvement across all levels of the company to ensure goals and commitments are understood and consistent.

89

Get responsibilities agreed upon by all team members, including review tasks that will ensure the project is meeting financial goals.

90

Create responsibility statements for each team member. Be clear and specific. Define ownership of results, not just effort.

91

Cross-functional collaboration means working together collaboratively and not staying in silos.

92

Keep assessing: What actions should the project take to achieve tightly-coupled cross-functional collaboration?

93

Take responsibility for cross-functional monitoring of business goals & profit drivers, even if it's not "officially" part of your job.

94

Always ask: "What do I need to do or communicate to help keep functional group's work and decisions aligned to the overall objectives?"

95

Use mini-plans proactively for different areas of functional work to thoroughly prepare for critical activities such as testing.

96

Use effective reviews to assess deliverables—are they meeting requirements? Are they still supporting the overall business goals?

97

The business savvy PM raises this question in every review: "Are we still meeting the most important goals of this project?"

98

Pay attention to "ilities":
Manufacturability, testability,
serviceability, reliability, certifiability.
All can impact or kill profits.

99

Plans and Reviews are not just about
the paperwork. They form a key
framework for a business-oriented
project process.

100

Plans and Reviews should not be rigid, be done with a bureaucratic "checkbox" mindset, or force meaningless work.

101

Mini-plans can be drafted & reviews can be held as soon as the information would help someone else's work or help avoid a later problem.

102

Plans and reviews should be held no later than the date past which you would have time to recover from any problems uncovered.

103

Keep asking the team, "Does this project have full, consistent, proactive cross-functional collaboration?"

104

Keep assessing: Is there a place where a key group is not adequately involved and needs to be involved right away?

105

Stay clear on the potential bottom-line impacts if cross-functional collaboration does not occur throughout the project.

106

Design reviews need to look at business objectives and profit targets as well as technical or other "content" aspects.

107

Use design reviews at different levels of design completion to keep ensuring profitable, working designs, and assess project impacts.

108

Early on a Preliminary Design Review should look at possible solutions in context of the business goals to help decide the approach.

109

Remember to review for issues that can affect profit such as sole source vendors, materials choices, design changes, & the "ilities."

110

Review proposed changes to avoid impacts to ROI and profitability. What is the cost-benefit of adding another feature?

111

Ask: Will adding this feature delay the project and thus when the company starts realizing its ROI? Is the benefit of it worth the delay?

112

Ask: Is this a risky feature that might require cost/time to correct and thus impact the profitability or return on this project?

113

Remember, documentation such as manuals—or lack thereof—can impact the business goals. Is the product usable? Is it maintainable?

114

A business savvy PM ensures the right project and functional documentation is done as part of the project.

115

Document manufacturing and test for accuracy and productivity.

116

Document maintenance and support for proficiency, accuracy, and productivity.

117

Document vendor requirements for accurate scoping & turnaround time, and support of profit goals.

118

Be ready to bring new employees "up-to-speed" quickly with an accurate understanding of the project, its goals, its deliverables.

119

Business savvy PMs set clear completion criteria to judge when the project is "done" from a business and profitability perspective.

120

Completion criteria for the project include what deliverables must be completed and what "done" means for those deliverables.

121

Completion criterion can also be set for specific project activities or interim project deliverables.

122

Feature-related criteria include feature completion, quality factors, customer test results, margin metrics, and manufacturing volume.

123

Business Savvy PMs establish and stand up for project completion criteria all the way through the project.

124

Continually ask during the project: Do I believe that adequate completion checks are in place to ensure we meet our goals?

125

The business savvy PM realizes that time is money, in multiple ways, & pays attention to personal & team priorities & productivity.

126

Staying focused to complete the project on time means monitoring personal work habits, project goals, and the company goals.

127

In monitoring where their time goes, each team member should ask, "Is this something the company depends on me and me alone to do?"

128

To avoid hits to the project, stay aware of who else is asking for team members' time. Are there priority conflicts to resolve?

129

Continually "carry the flag" for the business goals—build a business-aware mindset & personal decision-making into everyone's work.

130

Focus on results, and threats to results, at every step of the project. That is the business savvy PM's critical and impactful role.

131

Remember: During start-up, employ your own assessments & judgment to help define the business case & a project that will meet it.

132

Remember: During kickoff, drive the focus on business goals to find a solution worth doing from a business standpoint.

133

Remember: During execution, be business-minded and profit-aware, with continual checks of progress against the business goals.

134

Benefits to your career: Executives notice, value, and depend on business savvy project managers!

135

Executives value PMs who understand how to use project management to ensure the financial success of projects.

136

Executives value PMs who command the team's respect across functions.

137

Executives value PMs who know how to get the right things done and can communicate and make business-based project decisions.

138

The business savvy PM takes initiative and uses leadership skills to motivate others to get the full job done.

139

A business savvy PM is a bottom-line communicator who collaborates well across all functions to lead teams to the meaningful results.

140

Being a business savvy project manager is personally rewarding, opportunity building, and ultimately career enhancing!

About the Author

Cinda Voegtli is Founder and CEO of ProjectConnections.com, an online project management know-how source and service with over 300,000 members from around the world. She has over 20 years experience in project and portfolio management, process improvement consulting, and line engineering and company management. Over the years Cinda has consulted on project and portfolio management issues to a variety of companies and their unique projects, including Tyco Healthcare/Nellcor Mallinckrodt, Mobil Oil, Pacific Bell, Dow Chemical, NASA, Schlumberger, Aviron/MedImmune Vaccines, Hewlett Packard, and Lam Research. Cinda writes and speaks widely on practical real-world techniques for project and portfolio management that are organization-appropriate, easy to introduce for fast benefits, and adaptable for effective use on a variety of projects.

Other Books in the THiNKaha Series

The THiNKaha book series is for thinking adults who lack the time or desire to read long books, but want to improve themselves with knowledge of the most up-to-date subjects. THiNKaha is a leader in timely, cutting-edge books and mobile applications from relevant experts that provide valuable information in a fun, Twitter-brief format for a fast-paced world.

They are available online at http://thinkaha.com or at other online and physical bookstores.

1. *#BOOK TITLE tweet Book01:* 140 Bite-Sized Ideas for Compelling Article, Book, and Event Titles by Roger C. Parker

2. *#BUSINESS SAVVY PM tweet Book01:* Project Management Mindsets, Skills, and Tools for Generating Successful Business Results by Cinda Voegtli

3. *#COACHING tweet Book01:* 140 Bite-Sized Insights On Making A Difference Through Executive Coaching by Sterling Lanier

4. *#CONTENT MARKETING tweet Book01:* 140 Bite-Sized Ideas to Create and Market Compelling Content by Ambal Balakrishnan

5. *#CORPORATE CULTURE tweet Book01:* 140 Bite-Sized Ideas to Help You Create a High Performing, Values Aligned Workplace that Employees LOVE by S. Chris Edmonds

6. *#CROWDSOURCING tweet Book01:* 140 Bite-Sized Ideas to Leverage the Wisdom of the Crowd by Kiruba Shankar and Mitchell Levy

7. *#DEATHtweet Book01:* A Well-Lived Life through 140 Perspectives on Death and Its Teachings by Timothy Tosta

8. *#DEATH tweet Book02:* 140 Perspectives on Being a Supportive Witness to the End of Life by Timothy Tosta

9. *#DIVERSITYtweet Book01:* Embracing the Growing Diversity in Our World by Deepika Bajaj

10. *#DREAMtweet Book01:* Inspirational Nuggets of Wisdom from a Rock and Roll Guru to Help You Live Your Dreams by Joe Heuer

11. *#ENTRYLEVELtweet Book01:* Taking Your Career from Classroom to Cubicle by Heather R. Huhman

12. *#ENTRY LEVEL tweet Book02:* Inspiration for New Professionals by Christine Ruff and Lori Ruff

13. *#EXPERT EXCEL PROJECTS tweet:* Taking Your Excel Project From Start To Finish Like An Expert by Larry Moseley

14. *#IT OPERATIONS MANAGEMENT tweet Book01:* Managing Your IT Infrastructure in The Age of Complexity by Peter Spielvogel, Jon Haworth, Sonja Hickey

15. *#JOBSEARCHtweet Book01:* 140 Job Search Nuggets for Managing Your Career and Landing Your Dream Job by Barbara Safani

16. *#LEADERSHIPtweet Book01:* 140 Bite-Sized Ideas to Help You Become the Leader You Were Born to Be by Kevin Eikenberry

17. *#LEADS to SALES tweet Book01:* Creating Qualified Business Leads in the 21st Century by Jim McAvoy

18. *#LEAN SIX SIGMA tweet Book01:* Business Process Excellence for the Millennium by Dr. Shree R. Nanguneri

19. *#LEAN STARTUP tweet Book01:* 140 Insights for Building a Lean Startup! by Seymour Duncker

20. *#MILLENNIALtweet Book01:* 140 Bite-Sized Ideas for Managing the Millennials by Alexandra Levit

21. *#MOJOtweet:* 140 Bite-Sized Ideas on How to Get and Keep Your Mojo by Marshall Goldsmith

22. *#MY BRAND tweet Book01:* A Practical Approach to Building Your Personal Brand - 140 Characters at a Time by Laura Lowell

23. *#OPEN TEXTBOOK tweet Book01:* Driving the Awareness and Adoption of Open Textbooks by Sharyn Fitzpatrick

24. *#PARTNER tweet Book01:* 140 Bite-Sized Ideas for Succeeding in Your Partnerships by Chaitra Vedullapalli

25. *#PLAN to WIN tweet Book01:* Build Your Business thru Territory and Strategic Account Planning by Ron Snyder and Eric Doner

26. *#PRESENTATION tweet Book01:* 140 Ways to Present with Impact by Wayne Turmel

27. *#PRIVACY tweet Book01:* Addressing Privacy Concerns in the Day of Social Media by Lori Ruff

28. *#PROJECT MANAGEMENT tweet Book01:* 140 Powerful Bite-Sized Insights on Managing Projects by Guy Ralfe and Himanshu Jhamb

29. *#QUALITYtweet Book01:* 140 Bite-Sized Ideas to Deliver Quality in Every Project by Tanmay Vora

30. *#RISK MANAGEMENT tweet Book01:* Proactive Risk Management: Taming Alligators by Cinda Voegtli & Laura Erkeneff

31. *#SCRAPPY GENERAL MANAGEMENT tweet Book01:* Practical Practices for Magnificent Management Results by Michael Horton

32. *#SOCIAL MEDIA PR tweet Book01:* 140 Bite-Sized Ideas for Social Media Engagement by Janet Fouts

33. *#SOCIALMEDIA NONPROFIT tweet Book01:* 140 Bite-Sized Ideas for Nonprofit Social Media Engagement by Janet Fouts with Beth Kanter

34. *#SPORTS tweet Book01:* What I Learned from Coaches About Sports and Life by Ronnie Lott with Keith Potter

35. *#STANDARDS tweet Book01:* 140 Bite-Sized Ideas for Winning the Industry Standards Game by Karen Bartleson

36. *#TEAMWORK tweet Book01:* Lessons for Leading Organizational Teams to Success 140 Powerful Bite-Sized Insights on Lessons for Leading Teams to Success by Caroline G. Nicholl

37. *#THINKtweet Book01:* Bite-Sized Lessons for a Fast Paced World by Rajesh Setty

38. *#TOXINS tweet Book01:* 140 Easy Tips to Reduce Your Family's Exposure to Environmental Toxins by Laurel J. Standley Ph.D.

THiNK Continuity™ Training/Learning Program

THiNK Continuity™ delivers high-quality, cost-effective continuous learning in easy-to-understand, worthwhile, and digestible chunks. Fifteen minutes with a *THiNKaha®* book will allow the reader to have one or more "aha" moments. An hour and a half monthly with a THiNK Continuity program will allow the learner to have an opportunity to truly digest the topic being covered.

Offered online and/or in person, these engaging programs feature gurus (ours and yours) on such relevant topics as Leadership, Management, Sales, Marketing, Work-Life Balance, Project Management, Social Media and Networking, Presentation Skills, and other topics of your choosing. The "learning" audience, whether it is clients, employees or partners, can now experience high-quality learning that will enhance your brand value and empower your company as a thought leader. This program fits a real need where time and the high cost of developing custom content are no longer an option for every organization.

Just **THiNK**...

- **C**ontinuous Employee/Client/Prospect Learning
- **O**ngoing Thought Leadership Development
- **N**otable Experts Presenting on Relevant Topics
- **T**ime Your Attendees Can Afford – 15 min. to 2 hrs/mth
- **I**nformation Delivered in Digestible Chunks
- **N**ame the Topic – We Help You Provide Expert Best Practices
- **U**nderstand and Implement the Takeaways
- **I**nternal Expertise Shared Externally
- **T**raining/Prospecting Cost Decreases, Effectiveness Increases
- **Y**ou Win, They Win!

www.ingramcontent.com/pod-product-compliance
Lightning Source LLC
Chambersburg PA
CBHW071514200326
41519CB00019B/5945

* 9 7 8 1 6 1 6 9 9 0 6 2 6 *